# BRITISH LIVESTOCK LORRIES

BILL REID

AMBERLEY

# Acknowledgements

The majority of the photographs used in this book are from the author's camera, or from his collection amassed over thirty-five years. I would like to thank Allan Reid, Eddie Waugh, Ian Lawson, Wobbe Reitsma, Paul Heaneachon, David Morgan and Alec Syme for their assistance.

First published 2018

Amberley Publishing
The Hill, Stroud
Gloucestershire, GL5 4EP

www.amberley-books.com



ISBN 978 1 4456 7222 9 (print)
ISBN 978 1 4456 7223 6 (ebook)



British Library Cataloguing in Publication Data.
A catalogue record for this book is available from the British Library.

Origination by Amberley Publishing.
Printed in the UK.

# Introduction

Before the development of motor vehicles, livestock was driven on foot to local markets and slaughterhouses, which to some extent set the course of roads throughout the country. Some of these 'drove roads' still exist as today's public roads, taking the route of least resistance through physical features in the landscape and providing the modern road user with some interesting roads.

After the railway companies became established in the nineteenth century, bringing a well-connected network across the country, livestock transport in large numbers and over much longer distances became available. The individual railway companies amassed large numbers of livestock wagons and this continued well into the days of British Railways, facilitating livestock movement in a more efficient way and leaving the animals better presented for sale or slaughter. The railway network was reasonably efficient, but the livestock still had to be driven to and from railheads.

In the late nineteenth century motor vehicles were appearing and their potential to move goods became apparent. The early motorised transport tended to be adapted cars and did not make an impact on livestock transport until the 1920s, when larger and heavier lorries were being designed. These larger lorries were suitable for transporting decent numbers of livestock, which became a better proposition for farm-to-market movement.

By the middle of the twentieth century, road-going lorries were of a size and load capacity to take over much of the railway movement of livestock. These lorries, with some pulling trailers, were capable of long distances at a reasonable speed, making their means of transport more favourable to farmers and butchers alike. The livestock containers carried were of a basic design – sometimes being carpenter-built and heavy, and sometimes being built by the owners of the lorries and more lightweight.

The traditional local livestock lorry often had removable parts that extended the already fitted dropsides to make a high-sided body. There were no roof panels and weather protection, if any, was by a tied-down tarpaulin. In most parts of the country specialist body builders came to the fore, including Jennings of Sandbach, Tiverton Coachworks, Carmichaels of Worcester, Struthers of Jackton, A. C. Penman of Dumfries and Clark & Sutherland in Aberdeenshire, to name but a few. Many smaller agricultural engineering companies would make bespoke livestock bodies on request.

On that basis, for many years livestock bodies were generally of wooden construction or had steel framing clad with wood. Although easy to replace when damaged or rotten, the wooden parts were gradually phased out for alloy panels on steel frames in the 1970s, making the bodies more durable and easier to keep clean – a requirement that was becoming a high priority in livestock transport.

The types of lorries used for livestock transport were whatever the owner preferred. All makes of British-built lorries were used, with petrol or diesel engines. Two-, three- and four-axle types were used, depending on the loads expected to be carried and, to a certain extent, relating to other work that was to be carried out. Articulated livestock lorries only became a proposition in the latter part of the twentieth century, when new weight and length regulations favoured the type.

As the lorries became larger the livestock containers became longer, with increased carrying capacity. This was evident in a return to lorry and trailer outfits, which could have more load space than the already generous length afforded to articulated types. A gross weight increase to 44 tons for articulated types and so-called wagon and drags on a minimum of six axles saw an emergence of double-deck transporters for cattle and four decks for sheep. Effectively, this means of transport doubled up the amount of load and theoretically should have halved the amount of livestock lorries on the roads.

These large lorries were not well suited for local livestock movement and smaller lorries continued to be in demand, ranging from 7.5 tons gross to 26 tons on three axles. Furthermore, many farmers began to think they could run their own transport with a small lorry, or a trailer drawn by a 4x4 or tractor. This has led to the virtual demise of the local livestock haulier, or remover. Longer distance livestock removers are using specialist semi-trailers from Houghton-Parkhouse Ltd, Plowman and Pezzaioli, from Italy, among others.

The new breed of livestock lorry or trailer is fitted with hydraulic systems to raise and lower the decks, or floors. They also have automated ventilation systems and water providers, with lighting in all sections to comply with animal welfare conditions. Tanks are required to catch the effluent and the vehicles must be cleaned before every load.

Veterinary inspections can take place unannounced and all drivers must carry certificates showing they have a knowledge of animal welfare, all the while keeping a close record of the livestock moved in their lorries – a far cry from the drovers of old!

In the early days of motorised livestock transport the lorries were small and were used on a local basis. The railway companies developed large fleets, which could be used for many purposes, and the Great Western Railway had this small 1930 Thornycroft with demountable bodywork to service livestock traffic to and from railheads.

In those days, some livestock transporters preferred a fitted body, like on this 1930 Commer Invader. This allowed a much lower floor and gave easier access for the livestock load. The lorry would not have been suitable for any other use.

An early Bedford WLG dating from 1934 with a coachbuilt horsebox body and cab. It has been preserved and is seen finishing an HCVS London–Brighton Run on Madeira Drive, Brighton.

Another coachbuilt horsebox on a Leyland Cub chassis. It has an integral cab, not of Leyland manufacture, and in the 1930s this would have been an expensive vehicle.

An old picture of a 1936 AEC Matador Mk II, looking as if the lorry was new. It carries a rudimentary livestock body for cattle and sheep. The body can be demounted by slings, freeing the lorry for other work.

The post-Second World War years saw a number of specialist coachbuilders producing livestock containers. A. C. Penman of Dumfries, who were also Austin agents, built this low-floor body on an Austin K4. The upper or third sheep deck must have given the Austin a good load.

Road Services (Caledonian) was formed after the partial denationalisation of British Road Services. Caledonian provided all types of transport and this Albion CX is seen in their Abington Depot with a substantial lift-off livestock body. The Albion was the first vehicle registered by Kirkcudbright County Council in 1949.

A preserved Foden DG lorry with a substantially built wooden livestock body, which has been built for a shorter vehicle. The container is demountable and can be removed by the rings on the roof, or by a ramp and pole system, which will be seen later in the book. Note the method of securing the container, which was commonplace for many years.

In the early post-war years new lorries were difficult to obtain and many operators adopted a make-do or mend approach. Old bus chassis were converted to goods lorries and were a well-used basis for livestock transporters. This Bristol was a double-deck bus in a previous life.

New lorries became more available in the early 1950s, and this 1955 Foden FG with a matching livestock trailer looks new with the driver, or trailer boy, posing in front. Able to carry two normal lorry loads at a time, the law decreed that it needed a two-man crew to operate all the brakes.

One of the most popular lightweight lorries was the Bedford OL. It was used in all aspects of transport and this very fine preserved example represents Neil Bartlett's early business ventures.

New in 1950, this small Austin K8 is an ancestor of the ubiquitous 3.5-tonner of today. It was probably used originally by a Dunbartonshire farmer as local transport for his livestock and produce on market days.

Another Bedford OL, which has been preserved with a substantially built livestock body. It has been restored in the name of a prominent pedigree cattle breeder from southern England.

During the 1940s and '50s there was a vogue for heavyweight horseboxes built on bus or coach chassis. British Railways had several. This one, built on a Foden PVSC coach chassis, transported racehorses.

As the 1950s progressed a wide variety of British lorries were available. Seddon Diesel produced their Mk 5L as a long wheelbase lorry and it was readily taken up by livestock removers. This 1954 Mk 5L, based in Stranraer, had a varnished wood body by Penman of Dumfries.

A colour photograph of a varnished-wood-bodied Bedford OL, which was in identical colours as the above Seddon.

An Albion CX dating from 1949, completely restored by Mr David Struthers to the style in which his coachworks turned out livestock lorries in years gone. It has the Struthers characteristic luton box and a third sheep deck. The inappropriate age-related registration was allocated much to Mr Struthers's disgust!

A. C. Penman of Dumfries was a prolific coachbuilder and supplied many livestock lorries. This Austin Loadstar is demonstrating the Penman System for demountable bodies. The lorry was driven between the ramps until the poles through the body lifted it clear of the platform. Remounting the body was simply a reversal of the procedure.

One of the longest serving livestock lorries in Ayrshire was this Bedford S Type. It was new as a dairy products lorry before being acquired by a farmer who converted it to a tipper. New in 1955, this photograph was taken in the late 1980s. It still exists, alongside the two lorries that were to replace it!

Lightweight 5-ton lorries were used by many livestock removers, and the Morris Commercial FV5 was well liked. The cab had fairly easy access through rear-hinged doors, which were known by some as 'suicide doors'. The standard Morris four-cylinder petrol engine was a bit underpowered.

Long-distance livestock haulage was becoming the preserve of road transport in the 1950s and heavier lorries capable of pulling trailers or carrying larger livestock containers were more frequently seen. The twin-steer Foden FG was a good lorry for both purposes and this one demonstrates the third deck principle incorporated within the body. It would pull a similar trailer.

Another Bedford OL in preservation showing another aspect of livestock transport. In actual fact it was used to transport dead animals away from farms, which were winched up the rear ramps. The high sides kept the carcases out of view.

Articulated livestock lorries were uncommon in the 1950s; unless specially adapted, like this small wheeled step frame trailer, they were usually rather high and unstable, with a steep loading ramp. The type did not become popular until longer and heavier articulated lorries were allowed.

Traditional style wooden cattle containers were produced by a number of coachbuilders, usually for individual buyers rather than hauliers. This Thornycroft has such a body, which is finished in varnished wood. The paintwork of the cab has been done in a technique to match the wood grain.

Thomas McBride was a haulage contractor from the village of Crossmichael in southern Scotland. His AEC Mercury is an early example of a 12-ton gross model. The livestock container was built by Callander's of Minnigaff to fit the lorry, which is seen here at a local cattle show.

A new livestock lorry fully supplied by A. C. Penman of Dumfries. As Austin dealers, Penman could supply a complete lorry. It is likely that this Austin would have been presented at a Royal Highland Cattle Show in the Penman stand.

A rear view of a Penman livestock body, this time fitted to a Bedford TK and showing the ramps for the internal deckring. The upper ramp is split in two sections and would have been raised manually after the rear section had sheep in it.

A Morris FEK100 that was new in 1957. It had a 17-foot livestock body that was built entirely by joinery at the Maybole firm of Alexander Jack & Co., fully fitted with a removable sheep deck. The deck was removed, as many were, by manual handling when not required. Alexander Jack & Co. were well known as agricultural engineers.

Carmichael of Worcester was another well-known coachbuilder (which was better known for fire appliances). They did a respectable trade in livestock containers and this ERF KV lorry is a good example. The standard Carmichael body had a shallow angle at the rear end to ease the height of the floor and the angle of the loading ramp.

The Big Bedford, or S Type, was also a well-favoured lorry in livestock haulage. This one seems to have had a wheelbase extension to allow a longer platform. The livestock body is well built, presumably by a specialist, whose name is not known.

A 1962 ERF KV, operated by a farmer from Lanarkshire. Previously, it had been a Scottish & Newcastle brewery lorry. It ran for the farmer for many more years than it did as a beer lorry and carries a more modern steel and alloy container, which was possibly made by J. Shanks, of Denny.

Coachbuilt horseboxes have always been a tradition in equestrian circles and when the Big Bedford series was introduced, the PSV model with its longer wheelbase was a cheaper alternative to the heavier bus and coach chassis previously seen.

This Austin 5-tonner with a gross weight of just over 8 tons has a rather heavy-looking three-deck livestock body. The collapsible sides of the third deck look more than fit for the purpose. This body would have taken up a sizeable chunk of the lorry's payload even before it was loaded with livestock.

Another heavy livestock body on an AEC Mercury, but this lorry had a gross weight of 12 or 14 tons and a much better load capacity. Lloyds of Ludlow are still in business.

A horsebox built on an early 1960s Commer QX lorry chassis. It is an integral body, which has not incorporated much of the standard Commer cab. The driving position, to suit the normal cab, is well inboard of the body sides. It appears to be a Lambourne box.

Before the advent of heavier and longer articulated types, the four-axle, or eight-wheeler, was the king of the road. From a length point of view it had a similar capacity to two- and three-axle types, but was thought suitable to carry a double-deck livestock body for cattle. M. J. Parry used this early 1960s example on an AEC Mammoth Major Mk V.

Commer lorries were upgraded with a new and larger cab in 1962. This one operated by Lionel Pike Ltd had the Rootes Group two stroke engine, which produced a raucous note when working hard. The lorry is typical of many 7 tonners doing local livestock haulage.

The normal control Ford Trader was a late bonneted type of lorry produced when most lorry users were turning to forward control for greater load space, and few were seen as livestock lorries. This one, with a well-finished coachbuilt body, may have been used as a horsebox.

Registered new in 1960 by West Sussex C. C., this Bedford S Type was restored and seen attending a vintage lorry event in the Midlands about twenty-five years ago. It has a fully fitted livestock body with opening slats for two decks of sheep. This could be a Tiverton Coachworks body.

Under the coachwork is an Austin 5-tonner. The low build of the cab suggests a normal control chassis conversion. This would have been an expensive horsebox in its day.

Alloy livestock containers were beginning to appear in the 1960s and this Atkinson mounted box, thought to be a Clark & Sutherland product, would be an early example. The sheep decks can be seen folded against the interior sides, so that they did not need to be removed.

Glynn Swain, the MD of Swains of Stretton, uses this former horsebox on a Commer VA chassis as his motorhome while attending vintage lorry events. He has another older Commer QX horsebox that has been restored to the same standard.

This Austin FJK140 is in original condition and carries a Penman livestock box that was new on an older Austin, which this lorry replaced. The cattle box was to go on to an Albion lorry, which in turn replaced the Austin FJK140. All three lorries exist in preservation.

James Holmes (Greenock) Ltd had a small but fine fleet of livestock lorries that were finished in red with all the Scottish traditional touches. This AEC Mercury was the last new lorry in the fleet and carried a wooden livestock body built by Howie of Dunlop. Howie bodies were seen throughout Scotland, being the first choice for many farmers and haulage contractors.

An Atkinson Searcher, seen pulling out of the loading banks at the old Ayr Market. It has a capacity load of sheep in its Houghton-Parkhouse livestock body, which was built at Milnthorpe in Cumbria. The operators of the lorry were MacTaggart Bros, who are still very much in livestock haulage today.

New rules for lorries in 1964 brought about a redesign of Leyland Group types. This is a Leyland Super Comet with the new style Leyland Ergomatic cab, which came into preservation in the 1990s. It later carried a new Irish style livestock container, which is currently fitted out like a caravan for the comfort of the present owners, Mr and Mrs Henry Newall, while attending vintage events.

Irish livestock lorries tended to have open-top bodies, and this tremendous restoration of a Volvo FB86 shows the style to advantage. The matching A-frame drawbar trailer was a popular type for a long time, before the more manoeuvrable centre axle types made their appearance.

J&D Walker are long established livestock hauliers and their AEC Marshal has a wooden body with steel framing, which was a familiar style in the 1960s and '70s. The AEC has the Ergomatic cab like the Leyland Super Comet seen earlier.

As already noted, PSV chassis were considered good bases for horseboxes. Bedford produced a twin-steer coach chassis on small wheels to keep the coach floor to a low level. The chassis has been put to good use by a horsebox builder, allowing plenty of space for the horses and accommodation for the owners.

As seen before, the eight-wheeler chassis was considered as suitable as a double-deck cattle transporter. This well-known Atkinson Defender is even better suited, with a gross weight of 30 tons.

A different means of livestock transport; Catterall Pigeon Transport had this transporter built from a Ford PSV chassis for long-distance pigeon racing. A Duple coach cab was used with sleeping accommodation above.

The Hino truck was first seen in Europe in the Republic of Ireland, where they were assembled. This older example carries a traditional Irish livestock container that was fixed permanently to the lorry and with the usual open top.

The Volvo truck that revolutionised the UK haulage industry was the F86, which made its mark in articulated use. Some were found suitable as lorry and trailer outfits and as six- and eight-wheelers in livestock transport. This Bateman F86 has a locally built Houghton-Parkhouse body.

By the mid-1970s heavier articulated livestock lorries were appearing with longer trailers, affording more load capacity. This Walker Atkinson Defender has a livestock body with a third sheep deck. New rules were soon to come in, requiring the third deck to be an integral part of the body.

Michael Janes's amazing ERF A Series wagon and drag outfit. This lorry is one of only a few A Series that were built to drawbar specification. It was driven up from Somerset to take part in an Ayrshire Road Run for vintage lorries a few years ago and was the star of the show!

The A Series ERF was superseded by the B Series, with a new tilt cab. This one carries a three-deck body, which has the third deck as part of the whole rather than being in the open air as a temporary structure.

James Kinnear was a livestock haulier from Fife, in Scotland. This line-up of Scanias shows the third deck principle as a permanent conversion on the right, while the lorry in the centre has still to have its upper deck enclosed.

The Bedford TK range has always featured well in livestock haulage. This small horsebox is a rather basic alloy-panelled type, which has survived into preservation and may still be in use transporting recreational horses.

Since the introduction of multi-wheeler Volvos, the type has been popular in livestock haulage. This F7 model eight-wheeler has a double-deck Houghton-Parkhouse body, which is seen well filled with cattle.

Dennison Trucks were part of a large engineering company in Ireland and for some years produced heavy lorries. They were fitted with angular cabs by SISU, a Finnish lorry manufacturer, and various engine types. This one has a large Rolls-Royce Eagle diesel.

All types of lorry makes were used in livestock haulage. Bannatyne Motors, from the Isle of Arran, in the Firth of Clyde, used this FIAT 159, which in many ways was quite conventional, except for its unusual Trilex wheels.

In the east coast of Scotland there are some very large farms and estates. To cater for different haulage uses, a versatile lorry like this ERF B Series evolved. Built as an eight-wheel tipper, the body sides can be folded in or removed to allow it to run as a platform lorry. This one does extra duty as a livestock transporter with a removable body. It may have been able to pull a trailer.

Strathclyde Police was formerly the largest police force in Scotland and had many departments, including a mounted section. In 1979 this Ford horsebox was acquired. It is built on a PSV chassis, with accommodation for five horses plus the riders and grooms.

James Kinnear had one of the few Bedford TM six-wheelers in Scottish livestock haulage. It had a Bedford engine rather than a Detroit two stroke. It has to be said it is not looking its best in this picture at Stirling Kildean Market.

Henry Stewart was a long-term driver with John Craig Livestock Haulage until the untimely death of the owner. Henry struck out on his own using this Volvo F7 with a three-deck, privately built livestock box. It is seen parked at Stranraer harbour.

Allan D. Young was a user of double-deck livestock lorries on eight-wheel chassis and later with articulated semi-trailers. His Foden S10 of 1981 is pictured in the old Lanark Market. It has a double-deck container for cattle.

Certain smaller lorries were rarely seen as livestock vehicles. Karrier was a very scarce name around the markets. This one, seen in preservation, has been included to show what might have been. It has a small sized body more often seen on a Ford Transit type.

The Leyland Ergomatic cab lasted a long time, and this 1980 Lynx is part of the Leyland 500 range. The cab was raised some 4 inches to accommodate larger engines, but was basically the same as the original design. It is pictured here on a wintery day at Paisley Market.

A Continental style livestock lorry that has been imported into the UK. The Scania T112 retains left-hand drive and has a livestock body incorporating sleeping quarters above the bonneted cab. There is provision on the roof of the cab to carry fodder. (W. Reitsma)

Another small Bedford TK with a basic livestock body by Shanks of Denny. The container is a lift-off type using the ramp and pole method, which allows the lorry to do other work.

Whitecairn Transport took over the business of John Craig Livestock Haulage in Stranraer, changing the fleet colours, as seen on this Leyland Clydesdale. Craig had almost standardised their lorries on Leyland Clydesdales with Houghton-Parkhouse containers at the time of the sale. Craig and Whitecairn lorries were always kept in immaculate condition. (A. Syme)

Dennis lorries were another type not commonly seen in livestock haulage. This may have been due to their lack of numbers, as they were made to order vehicles. This one, with a Perkins engine, was the standard lorry among fire appliances and refuse collectors.

M&R Knight seem to have been happy with Dennis lorries as they were also running this newer version known as a Delta. The cab was designed to fit all the heavy chassis made by Dennis at the time. The lorry looks new, with a new Houghton-Parkhouse container.

MAN trucks collaborated with Volkswagen to produce a small lorry at 7.5 tons gross, which at the time could be driven without the need for an LGV driving licence. A drawback with this size of lorry was that they could be easily overloaded with large livestock.

A Bedford TL, which was to be the successor of the Bedford TK, with a Tiverton coachworks livestock container. A restored lorry, this represents many of this type and the coachwork used in the south of England.

An example of the more common Bedford TK with a Houghton-Parkhouse container. The TK cab was quite long, making for a large luton box above it. Note the Q registration. The Bedford is probably an ex-MoD lorry.

With changes in market forces, farmers and dealers were travelling further to buy livestock; consequently, larger lorries made sense when longer distances were being covered. A. Redpath of St Boswells used this MAN 19.321 with a single-deck, 40-foot trailer for cattle or as a third-decker for sheep.

Every September the town of Kelso hosts important sheep sales and livestock lorries from all parts of the country arrive at the town. Robert Frame's Volvo F10 has the third deck erected but was probably used for carrying fodder rather than sheep. The container had been adapted to fit the Volvo, having been used previously on a lorry with a much lower cab.

James Roberts ran no more than two livestock lorries at a time. His 1984 ERF C40 served him well with a Houghton-Parkhouse container on a step-frame trailer. The ERF is seen in the old Ayr Market, which is now the site of a Morrison's supermarket.

Parked in Stirling Market awaiting a load is a Wm Armstrong Leyland Roadtrain of 1984 vintage. The double-deck trailer has taken some damage from trees, but appearances have been kept up with a repaint. Armstrong is still operating in livestock haulage. (A. Reid)

The Dourie Farming Company from Port William in southern Scotland used this enclosed three-deck Houghton-Parkhouse on a Foden S10 chassis during the 1980s for the carriage of cattle and pigs. The heavier uprights on the body sides guided the decks when being lifted and lowered.

A DAF 2100, which may have been an artic unit converted to a long wheelbase lorry, with a third axle added. The open-air construction of the third deck is highlighted by the sunlight on this pleasant day in Castle Douglas Market.

Another Foden S10 lorry. It is pulling a matching trailer and would be capable of running at 38 tons gross with its five axles. The builder of the livestock bodies is unknown.

One of Allan D. Young's articulated livestock lorries, parked in Stirling Market. There is still a height problem dictated by the tractor unit fifth wheel, and double-deck artics used a step-frame trailer to keep the overall height to a minimum.

After the Seddon and Atkinson merger, new types emerged. This is a Seddon-Atkinson 301, which was designed as a lightweight six- or eight-wheeler. R&W Young operated from Yetholm, in Roxburghshire. The container is a step toward the integral three-deck type, but it is not roofed over, which eventually came to pass.

The DAF 2500 was designed as a lightweight articulated unit, and this one, with a bespoke horsebox, may have been such a lorry extended to take the large body. It had brought horses to Ayr Agricultural Show, and the hunting riders can be seen reflected on the side.

Luce Bay Plant Hire may be a strange name to find on a livestock lorry. The company was part of a large farming and cattle-dealing enterprise in the same name. The lorry, a Leyland Roadtrain, was an extended tractor unit with a third-axle conversion. It pulled an equal sized trailer.

Campbell of Oxton has been a well-known livestock haulier over the years and has run a varying sized fleet of rigids and articulated types. Their 1985 Volvo F12 and a similar Volvo F10 were in Castle Douglas Market awaiting loads. Again, the open third deck is highlighted.

Found visiting Dumfries Market was J. A. Hewitt's ERF C Series, which was collecting cattle for the firm's butchery business. Jewitt's business is based in Sunderland, so the lorry had quite a cross-country journey ahead of it.

Oliver Bros, from Denholm in the Scottish Borders, had this Mercedes-Benz 1617 lorry uprated with a third or pusher axle. This was a sensible conversion to avoid overloading the axles. The livestock container has been kept outside and bleached white by the sun.

A Seddon-Atkinson 301 in service with Redpath of Wooler, which had discharged its load of livestock at Kilmarnock Abattoir. Redpath had quite a striking red livery, which stood out well in traffic.

A lorry operated by a farmer and cattle dealer was this medium-weight IVECO 135. Running at 13.5 tons gross weight, it would be useful to the owner who did not use it for other haulage. It carries a Shanks of Denny container.

Fully laden with sheep on all three decks of the Haughton-Parkhouse trailer, this DAF 2800 is waiting to pull out of Carlisle Market. It may have a long journey ahead of it as McCallum is based in the north of Scotland. Note how the top deck load is exposed. Sheep falling out after clambering on others was not unheard off.

Another large double-deck livestock transporter based on a Foden eight-wheeler. The container looks like a Paton construction, from Aberdeenshire. The firm of Peter G. Barclay changed hands and operates today as Barclay's Transport.

An ERF E Series, all polished up and attending a truck event in Carnforth Truckstop some years ago. The Houghton-Parkhouse container has been given the same treatment, but the Ellison lorries have always been well kept, so it maybe it did not take much bulling-up.

David Sloan used this Scania 82 as a part-time livestock haulier. The container could be lifted off with jacks using the attachments on the side. Note the steering angle, which was a boon for Scania drivers reversing trailers.

A medium-weight Ford Cargo used by an Ayrshire farmer between his farm and local markets. The Shanks container had been used with a lower-cabbed lorry and was suitably modified to fit.

John C. Eaglesham was an owner/driver who worked into the South Ayrshire and Galloway markets. John's lorries were always immaculately maintained and he had a succession of Bedford TKs that were identical until this Bedford TL was acquired. It was his last lorry.

Robert McCaig is a farmer and agricultural contractor. For a time he ran heavy livestock haulage, and this Volvo FL10 was his first lorry. Houghton-Parkhouse bodies are fitted to the lorry and trailer.

Seen at Carlisle Market is this Scania wagon and trailer combination, covered in the muck and detritus from wintertime roads. The security of the trailer container leaves a lot to be desired!

A. Redpath of St Boswells had this well-used Volvo FL7 on livestock work at Stirling Market at the time of a large sheep sale. The upper deck has been left erected in anticipation of a good return load of sheep. The container security, by straps, looks flimsy, but it did work.

In for a large load at Ayr's new market, T. Bell's ERF is equipped to run at 38 tons on six axles, allowing for a much reduced road tax charge at the time. T. Bell has since ceased trading.

James Roberts' second lorry was this MAN 19.340 that pulled a low, single-deck livestock container, which gave it the impression of being very long. It ran between markets and local abattoirs, where access was restricted by height.

An ERF E10 operated by a farming company near Perth. The lorry had the versatility seen before of being a tipper that could be converted to a platform lorry, and was therefore able to take the demountable livestock container.

T. McBride Transport had this heavyweight DAF 3600 running with a similar sized trailer from his base near Castle Douglas. Both containers are of Houghton-Parkhouse make, which were very popular in southern Scotland.

William Milven & Sons operated a livestock haulage business from the family farm, Kirkland of Longcastle, in Wigtownshire. One of their smaller lorries was this ERF E6, which originated in Cumbria and carries the ubiquitous Cumbria-made Houghton-Parkhouse container.

Development of articulated livestock trailers saw them becoming integral units, with the structure taking much of the load. This allowed for a lower floor and centre of gravity. Cattle and sheep lifts were installed for the upper decks. The Volvo F16 would be well in charge of the trailer.

A similar livestock trailer operated by L. E. Jones from North Wales but pulled by a lower powered Volvo F10.

R. G. Barr Ltd is a livestock and fodder transporter. This Scania 93M is in retirement and is kept as a show truck. It was taking part in an Ayrshire Road Run when seen in the picturesque Galloway village of Portpatrick.

The livestock lorries of Henry Thomson are a well-known sight around the country, and have been for many years. Their 1986 ERF E14 was in the wash at Ayr Market with, unusually, a trailer from another haulier. It may have been a brought-in trailer about to be given the Thomson treatment.

The Scottish Borders is prime sheep and cattle country, with numerous market towns spread across the area. Oliver's Transport is a haulage and livestock removal business that has prospered operating from Kelso. Their distinctive blue, white and red livery stands out well on this Scania.

James Craig Ltd was a latecomer to livestock haulage, having taken over the business of Peter Strain, in Newton Stewart. Craig ran a diverse fleet of livestock lorries, in articulation and as lorry and trailer combinations. This Volvo FL10 was converted from an artic unit to a drawbar lorry.

An ERF E10, which was operated by a long established livestock haulier from Dumfriesshire. William Johnstone kept a fleet of ERFs and Fodens busy in southern Scotland. The ERF is being washed at Ayr, and has one of the popular Houghton-Parkhouse livestock boxes.

Up from Yorkshire at Carlisle Market, the Foden 4400 of R. Brownbridge is unusual, being a two-axle type. It has a 400 bhp engine, which would be well on top of the job as it could only gross at 32 tons. Note the through loading method from the trailer to the lorry container.

Also in Carlisle Market was this medium-weight Renault lorry. It was operated by a local farmer around the markets in the area.

J&D Walker had this impressive Seddon-Atkinson Strato as a wagon and drag livestock outfit. It has a traditional three-deck container made by Paton of Netherlee, in Aberdeenshire. The short trailer has a much older container with an open third deck and through loading front doors.

H&D Fraser & Sons operated from Conon Bridge, north of Inverness, on general and livestock haulage. This ERF E6 was restored by a member of the Fraser family and is a regular entry in vintage events across Scotland, displaying the fleet livery and the company mission statement.

Full of sheep on all four decks, this MAN 17.292 is all ready to roll. This is the sort of picture that is anathema to the animal rights movement. William Gibson is a long-established livestock haulier, with plenty of experience of shifting loads like this, safely and humanely.

A fine combination of a Volvo F12 and trailer, looking in excellent condition at a cattle market in southern England. The Volvo has a cab-top sleeper cab, which was a Volvo option and known as the Eurotrotter. Some drivers thought them claustrophobic and difficult to get into. (P. Crang)

R&W Young was another Scottish Borders livestock remover. Their Seddon-Atkinson Strato artic unit shows off the company's traditional livery, against their trademark light brown livestock containers. R&W Young was taken over some years ago, by a nearby competitor. (P. Crang)

The Stranraer area of Scotland saw changes in the companies operating livestock haulage. E. Copeland is a coal merchant who branched into livestock and general haulage. Their tidy looking Volvo FL6 with a Paton livestock container is being washed at Ayr Market.

R. W. Stewart's livestock lorries have been a familiar sight around the markets of Scotland for many years. Their Scania 113 eight-wheeler with a three-deck livestock container must have been one of the largest rigid livestock lorries on the road.

A. Kerr & Co. were farmers and cattle dealers from Dundrennan in southern Scotland. The business developed from local livestock haulage to long-distance work with double-deck artic trailers, as seen with their Volvo F12 on the cobbles of the old Ayr Market.

Fodens did not seem to be popular lorries with livestock hauliers and were only found in penny numbers. R. H. Gooding ran this Foden 2210, with a fully enclosed three-deck container, in the West Country.

David M. Reid & Son ran a small fleet of lorries with demountable livestock containers in and out of the markets of south-west Scotland. This smart MAN 17.190 has a Shanks of Denny livestock box and carries the traditional green and red colours that were established in 1949.

A smart horsebox body on a Scania 93 chassis. Scania's cabs were, by this time, intended to tilt for maintenance and repair to the engine, but the cab front and side are integral with the body. The cab floor must then have been removable to get at the engine.

Chris Waite has been involved in long-distance livestock haulage for many years. His Renault Magnum was an unusual type on this type of work. The Magnum cab was well ahead of its time in 1991 and it took other truck manufacturers a long time to come up with equivalents.

Another scene on the cobbles of the old Ayr Market with the main part of the picture showing James Kerr's Scania 143 being loaded. In the background is a James Craig lorry, which has already been mentioned. James Craig Ltd was also the owner of the market.

An impressive ERF E14 with a Houghton-Parkhouse trailer at the annual ram sale in Kelso. The trailer is a four-deck type and would not be filled to capacity as each ram would probably be segregated, being worth large sums of money. Dalkeith Transport withdrew from livestock haulage.

Loaded with cattle from Ayr Market, this Scania 113 was travelling to an abattoir near Stirling in what would have been a fairly easy journey for the animals. Both third decks of the combination have been enclosed, as was the requirements at the time.

Fred Davidson started in road haulage in the 1920s in Dumfriesshire. The business is still being operated by the Davidson family. Livestock and general haulage is undertaken locally and over long distances. The present-day fleet is mostly made up of DAFs, like this 85 being prepared for another load. Lorry and trailer combinations as well as articulated types are operated today.

Mason Bros is a long-term haulier from south Lincolnshire, with a large fleet running on general haulage. Their tidy Scania 113 Streamline was attending the Royal Highland Agricultural Show, near Edinburgh, and waiting to load up for the trip home.

Also attending the Royal Highland Agricultural Show was this rare Foden 3350 livestock lorry with a Paton container. The container has a rolled up sheet, which would be used to cover the third deck when erected. With a 350 bhp engine, the Foden probably pulled a trailer.

The Renault Magnum was rare in livestock transport, and more so as a horsebox. This good looking lorry has a basic box as the horsebox, but it would be well fitted-out for the horses and the attendants.

Pictured in the wash bay at Ayr Market, this Les Oliver Scania 144 epitomises the long-haul, double-deck cattle transporter. The V8 engine has the power to sustain fast running times, with plenty of torque to maintain speed on hills.

Four-axle lorries benefited from a slight increase of 2 tons in their gross weigh in the 1990s to 32 tons. John Gyle, an Aberdeenshire farmer, had this Volvo FM12 tipper, which was easily converted to a platform lorry to carry the double-deck cattle container.

Hume's Transport, from Diss, put this four-axle Scania 124 into service in the late 1990s. It is unusual in having a fixed livestock container, meaning it could only be used in this type of haulage. The slightly mismatched trailer would allow the combination to run at 44 tons gross.

The DAF 65 was a popular all-round lorry at 17 tons gross. An Ayrshire farmer used this one with a two-deck body. The second deck can be seen folded against the sides while the lorry is being prepared for a wash at Stirling Market.

The Stamper name is well known in Cumbria in road haulage circles. R. J. Stamper has been in livestock haulage for many years. The Scania 112 has been restored and is shown at vintage events across northern England. Note the mural depicting cattle in the Lake District.

The Walker family from Strathkinness, in Fife, continued the business of livestock haulage with this good looking Scania 124, seen at the Kelso ram sales. It has a fully enclosed three-deck body, which would double up as a cattle transporter, with the second deck folded against the sides.

Another of the Stamper livestock lorries with a Scania 124 artic unit. The Scania is pulling a four-deck Houghton-Parkhouse sheep trailer, which could also be used as a two-deck cattle transporter. The Stamper fleet is always kept in excellent condition, depending on loads and the weather.

Long-distance livestock haulage is the preserve of the high-powered artic units. This B. McCann Volvo FH16 with 520 bhp would have plenty of power for its Continental trips.

A DAF 65, new in the Republic of Ireland in 2000, carries the traditional Irish style of open-topped livestock body. No provision appears to have been made to attach a tarpaulin cover. The body, built by an engineering company, is fixed to the chassis, rather than a demountable.

The Ayr Racecourse in south-western Scotland has several prestige horse races through the racing calendar. Horseboxes come to Ayr from various parts of the country, and this Mercedes-Benz from North Yorkshire was a nice example. It has retained the standard cab with attendants' accommodation in the separate body.

A close-up view of an Oliver of Kelso Scania 114 wagon and drag outfit at Bellfield Roundabout, above the A77 in Ayrshire. The lorry and matching trailer were heading for Kilmarnock Abattoir with fatstock cattle. (P. Heaneachon)

Lawsons' Haulage from Cockermouth is another well-known livestock haulier, although most of their work is in general haulage. The DAF 95XF was a popular lorry for this work, with plenty of power and good driver accommodation.

A line-up of Woodhead Bros livestock lorries, with a Foden Alpha nearest the camera. An ERF and a Volvo can also be seen, showing a certain diversity of choice of types in the fleet during the early 2000s. Generally, the Woodhead fleet of the present day comprises Volvo artic units.

The Longhorn family has been in haulage for many years and have a striking fleet of livestock lorries. This DAF 95XF with a 580 bhp engine and Houghton-Parkhouse trailer epitomises the modern fleet.

A well turned out DAF 85CF, with another Houghton-Parkhouse double-deck trailer which could also accommodate four decks of sheep. George Gatherer ran a small fleet servicing the markets of south-western Scotland.

The Dundas Chemical Co. seems an unlikely business to be in livestock haulage. The website tells the story, though. The Mercedes-Benz Atego is used for the collection of dead livestock across the country. (E. Waugh)

Parked in Ayr Market was this Scania 144 from the Northern Irish Laverty fleet. For reasons not known, it is pulling a Wm Armstrong double-deck trailer.

Another E. Copland lorry with a demountable livestock body, which is seen unloading cattle into Ayr Market on a sale day. The Volvo FM, with a third, or tag-axle, is in excellent condition.

L. E. Jones, of Ruthin, in North Wales, runs a number of double-deck livestock lorries within its otherwise general haulage fleet. The distinctive colour scheme stands out in markets and on motorways alike.

The haulage activities of J. Cooper & Co. dictate that their lorries and trailer have to be versatile, so they carry three-deck, demountable livestock containers. Load space is maximised for hay and straw haulage by having a non-sleeper cab.

Eric Gillie is a prominent horse transporter and well known for his fleet of Scania four- and six-wheel premium horseboxes. This excellently turned out Oakley Supreme on a Scania P310 chassis shows the high standard of the fleet.

Thomson of Sauchen went through some diversification into separate companies involving tipper and concrete haulage. Eric Thomson maintained livestock haulage and his DAF 105XF is seen awaiting a load in a cattle market. The traditional family livery was retained.

W. H. McWilliams had its origins as a local carrier and livestock remover in the small village of New Galloway, in southern Scotland. It recent times the business has grown considerably with large articulated livestock lorries such as this Scania R440.

Gradually being loaded on its four sheep decks, this Wm Armstrong Scania was about to take a full load of sheep away from Ayr Market to a Manchester abattoir. Note the many ventilation fans on the side of the trailer.

A very smart example of a Gordon Gilder Scania and livestock trailer. The Gilder family are well-known livestock hauliers from Gloucestershire. The Scania R580 looks new.

Seen at a North East Truckers show a few years ago was this small MAN TG-M with a Houghton-Parkhouse livestock container. It looks much lighter than its 15-ton gross weight.

Double-deck livestock semi-trailers are ideal for moving large amounts of animals, but have limitations in some of the older farm steadings. A. Jamieson & Sons use this short trailer to get into the less accessible farms in their area.

Skeldon's Transport had this modern style livestock container on a DAF 85CF on display at the 2010 Scottish Truckfest. The container is a fully enclosed three-decker, with hydraulic lifting of the upper decks. A degree of streamlining is evident on the container to match the cab air management.

Like many haulage companies, R. W. Stewart has stepped away from the traditional full livery and is using a modern style on a white cab. Their Scania R730 is a powerful artic unit and is capable of pulling the Plowman trailer, still in the traditional style.

Paterson Transport, from Wales, uses this Volvo FH with 540 bhp to pull their double-deck trailer on runs to English markets and abattoirs. It was pictured on a rest break in the Welsh mountains. (D. L. Morgan)

Looking quite compact with its fixed-body Plowman livestock container, this Scania R520 is capable of pulling a trailer. The cab and air management blend in neatly with the container. With 520 bhp, the Scania will be a nice drive without the trailer.

One of the 'Tag Team'. MacTaggart Bros have been involved in livestock haulage since 1971. The blue and red fleet is well known across the UK with their double-deck livestock trailers and curtainsiders on general haulage. This Scania was being loaded with sheep at Ayr Market.

A brand new Plowman outfit for Mansefield-based Bowring Transport, based on a Mercedes-Benz Actros chassis and appropriately entitled the 'Sherwood Hog Haulier'. (Plowman)

Few livestock lorries receive a great degree of customisation in the UK. J. Hulston's Scania R470 has been tastefully done with traditional Scottish tartan and thistles, along with a golden eagle presiding over a Scottish rural scene.

Another Laverty Scania from Northern Ireland, this time with a matching double-deck livestock trailer. The Scania R580 looks fairly new in this photograph from the summer of 2017.

Eric Thomson, of Aberdeenshire, has this Scania R580, which is seen with a Plowman trailer in Ayr Market, regularly collecting loads of cattle for abattoirs around the country.

Seen here with a much smaller capacity trailer is J. T. Orr's DAF 85CF. The trailer is the modern type with centre axles and a long drawbar, making for easier manoeuvrability.

A clean, well-presented Scania R480 fitted for drawbar work and operated by James Johnstone with a modern, fully enclosed livestock container in the loading banks of Ayr Market.

Cree Valley Transport, from Newton Stewart, is a small livestock haulage company that operates this DAF 75CF with a three-deck container. Their other lorry is also a DAF, with a double-deck Houghton-Parkhouse trailer.

Stranraer Farming Services operate this Foden Alpha, which pulls a Plowman livestock trailer. The Foden Alpha would be among the last built, in 2005, before production of Foden lorries ceased.

Italian livestock trailer makers Pezzaioli found favour initially in Ireland, and later in the UK. This one demonstrates the ability to lift the roof for easier loading and cleaning.

A line-up of part of the L. E. Jones fleet, with four double-deck articulated livestock transporters and Scania R580 artic units. The Jones livery has changed over the years to this striking, almost all-over orange style.

Various types of livestock lorries were commissioned for special uses. This small lorry and trailer combination, based on a Mercedes-Benz Atego 7.5-tonner, was used by a sheep and calf dealer who needed more floor space than weight capacity. (N. Scaife)

A new breed of eight-wheeler lorries developed recently with three axles to the rear are known as Tridems. George's Transport, in the West Country, use this 500 bhp Volvo FH with such an axle setup. To assist in manoeuvrability, the two outer rear axles steer. The single-drive axle might have limitations in messy farmyards.

A horsebox with a large caravan in tow. The MAN TG-M chassis carries a custom-built horsebox with no apparent attendants' accommodation, hence the caravan in tow. The large area above the cab has no windows and may be a storage compartment.

Not the ideal combination, but in order to fulfil obligations the owner of the livestock container borrowed the long single-axle trailer and had it coupled to the Scania 111. A short-term measure, it worked well.

Something completely different! Beedie Bros use large-capacity double-deck livestock artics, but getting them into older farms was difficult, if not impossible. The solution? Fit a demountable Jak-A-Box frame with a livestock body to the fifth wheel of a three-axle artic unit. It could go anywhere and was available as a proper artic unit with the box set aside. (Ian Lawson)

Perhaps not the newest of horseboxes, this Ford Cargo with Oakley bodywork was in excellent condition at the 2008 Great Dorset Steam Fair, and could have made an excellent entry in the vintage commercials section of the show.

Your author was brought up among livestock haulage and the last lorry his father drove was a Leyland Super Comet, as immortalised in this fine painting by Keith Hudson in 1995. The painting was done from two photographs – one showing the Leyland, and one showing a John Eaglesham Bedford, with the Leyland superimposing the Bedford.

Such was the interest generated by the painting that Eddie Waugh, a talented truck model maker, produced a model of the Leyland in approximately 1/43 scale – a fitting piece of memorabilia for the author.